Lincoln in General McClelland's Tent

Recipes,
Cooking Practices and Foods
from the Era of the
Civil War

Copyright© 1992 Jim Long

Long Creek Herbs
P.O. Box 127
Blue Eye, Missouri 65611
www.Longcreekherbs.com

Printed in the United States by Paul's Printing, Reeds Spring, MO
Second printing, June, 1993
Third printing, August, 1995
Fourth printing, September, 1998

ISBN #1-889791-06-7

⧼ Preface ⧽

The axiom that an army travels on its stomach has been proven again and again. Hungry troops and starving soldiers could barely think of going into battle when the rumble of their stomachs rang louder in their ears than the approaching cannon.

The armies of both sides in the War Between the States set up excellent lines of supplies initially. They trained and outfitted large numbers of cooks and their helpers and they fed enormous numbers of hungry men during the years of the war.

When we think of war we see in our mind's eye one army set against another. The reality of war most often was that small units had specific duties away from the main army. Small skirmishes were common. Ten or twenty men would be sent off to forage for supplies or to go on raids. Still smaller groups and individuals were spies or scouts, checking out the countryside in advance of the larger army. These small units had to rely on the food they carried, the game they killed, or upon the goodness and kindness of strangers to feed them.

Some soldiers knew how to cook, having come from rural areas and taken care of themselves before the war. Some learned recipes from watching mothers, grandmothers, grandfathers, aunts or wives cook at home. Many soldiers had had to fend for themselves before the war and were already fairly knowledgeable cooks. Others, out of necessity, learned simple cooking basics quickly in order to survive.

These recipes are from various collections and sources, concentrating upon how a soldier survived in smaller groups. There are entries for what soldiers might be offered when passing through friendly territory, as well as what they might find hidden in cellars or attics when on forays into unfriendly lands. There are recipes for making simple meals outdoors, as that's how the men lived for extended periods. My own great-grandfather, Alexander Sumpter, spent four years, living outdoors and moving with the Confederate Army with the 29th Tennessee Infantry.

Included here, also, are recipes from a few notable figures of the War, as well as a some instructions and recipes for caring for horses and keeping boots dry. The life of the common soldier was at once simple and difficult, dangerous and tiring, but regardless of his situation, he had to eat to live. Recorded here are a few of the recipes and foods of that time.

1

Red Eye Gravy

The story is that Andrew Jackson (1767-1845) the 7th President of the United States, was indirectly responsible for this recipe that became standard fare on many Southern tables.

General Jackson, while in the field with his men, called his cook over to where he sat to discuss what food he wanted prepared that day. The cook had been drinking hard the night before and was suffering from a bleary-eyed, staggering hangover. The General, seeing that the conversation wouldn't progress far in the state the cook was in, ordered him to bring his breakfast of ham with gravy "as red as your eyes." Red Eye gravy was born on the spot. The cook poured hot coffee into the dark ham juices and brownings from frying the ham and served it up. Men in the unit began to call for "cook's red-eye gravy" and the recipe became standard fare.

Use a cast iron skillet and fry several slices of country cured ham. Add some fat trimmings if available - enough to make about 1/2 cup of grease. (You can substitute lard or shortening if the ham isn't fat enough to produce the grease, or even fry the ham in bacon grease).

Fry the ham on a fairly hot fire, searing the ham well on both sides and leaving some of the ham "brownings" in the bottom. Don't burn it, but sear it well. Remove the ham from the skillet and set aside. Carefully pour in 1/2 cup strong black coffee and add a cup of water. Bring to a boil and serve over ham, biscuits, grits or potatoes. The gravy separates, leaving "red eyes" on the plate. Some families thicken the gravy slightly, using a tablespoon of flour dissolved in the last addition of water, but this doesn't leave good "red-eyes" in the gravy.

Jerky

Jerky was a staple food of many of the Indian tribes and frontiersmen quickly adopted this method of preserving meat. Jerky can be made from any meat, but was commonly made from larger game animals that were too large to consume quickly. Venison, bear, elk, buffalo, even goose and prairie chicken, could all be preserved in this way when the supply was bountiful.

There are various methods for making jerky, but the most common was to slice up the meat in thin slices, laying the slices out on racks made of green saplings. The racks were built into a smoking pit.

The pit would be about 3 feet by 3 feet, dug into an embankment so that the lower side could have a trench into it to add firewood. Wood would be stacked around the top of the pit, making a shelter and holding in the heat and slowing the escape of the smoke.

A small, smoldering fire would be built in the bottom of the pit, using some green and some dry wood. The object was to dry and smoke the slices of meat, not to cook it.

The tire would be tended and kept just barely going for a full day, or until the meat was about like shoe leather. The jerky was then ready for storage and could be taken on extended journeys.

Often jerky was cut up and boiled for an hour or more in a soup or stew. Wild onions, potatoes and any other available vegetables would be added, along with lard or animal suet. Without benefit of cooking implements or time, the jerky would sustain a man for days if necessary, just by biting off and chewing pieces as he walked.

Indians in drier regions of the country would hang the fresh meat on wooden racks in the sun. The air and hot sun would dry the meat quickly (this doesn't work in areas where there are flies, nor where humidity slows the drying process). Several days would be required to make jerky this way and it was taken inside at night or covered to avoid dew fall, then returned to the drying racks the following day. The smoking method, however, was probably simpler and had the added benefit of giving flavor to the meat.

Pemmican

Many frontiersmen of the early- to mid-1800s, well before the Civil War, learned from Native Americans the method for making pemmican. This was a highly nutritious food intended for long journeys where cooking implements might not be available, or for use on extended trips where time did not allow for stopping to hunt or cook. Some Indian tribes had been relying on pemmican as a high-energy travel food for centuries and lots of soldiers from rural areas knew about, and made use of pemmican.

Pemmican consists of any dried meat or jerky, pounded well and mixed with dried berries and nuts and held together with tallow.

A Typical mixture would be buffalo or venison jerky, pounded well and mixed with dried huckleberries, a little parched corn and roasted acorns, also pounded or ground. This was mixed together into a stiff paste with buffalo grease. The pemmican was dried in pouches or containers then carried in the pockets during travel. A man could sustain himself for long periods on this diet as it contained a well-balanced mixture of protein, fruit sugars and high-energy fats. Backwoods soldiers, those men from rural areas, knew pemmican well, as did the Native Americans who fought in this war.

Stonewall Jackson's Pork Ribs

Thomas Jonathan Jackson (1824-1863) was a fierce soldier who, once he planned his strategy, left no room for retreat. That stubborn dedication earned him the name of "Stonewall."

General Jackson was respected by Union and Confederate forces alike. He was deeply religious, respectful of the people around him, and a good cook as well. He was interested in the foods he ate and saw to it that his soldiers had the best foods that were available to him in a time of war.

His own concoction for barbecued ribs is as follows:

Start with pork spare ribs with plenty of meat on them. In a large stoneware jar or bowl put in about 8 cups of plain water. Mix in 1 level teaspoon of salt, 3/4 teaspoon of saltpeter, eight whole cloves, crushed, 1/8 teaspoon black pepper. Add the pork ribs, cut apart, mixing all well. Cover the jar and let stand for three days in a cool place, mixing every morning.

Drain the ribs at the end of the third day and place them on a raised wire rack over medium-hot coals to cook slowly. Jackson was said to have liked these cooked slowly in the oven, baked in a roaster, as well.

U. S. Senate Bean Soup

There's ample controversy about this recipe. Some credit the origin of Senate Bean Soup, that famous, thick bean soup that's been served in the Senate dining room for generations, to Senator Knutson of Minnseota. Other references list Jean Debruyn, a Belgian, as the author of the recipe. The Senate, regardless of the soup's author, was an important player in the Civil War.

Mr. Debruyn was an anti-slaver, expressing his views that no man had the right to enslave another. He did his best to arouse public opinion in Europe to put a stop to slave traders who kidnapped native peoples from the African coast and sold them in America. Arabs, who were deeply involved in slave trading, murdered him to silence his opinions and served to silence those who were around him.

Here's his recipe, which became

U.S. Senate Bean Soup

2 cups beans, soaked overnight in water, then drained
2 cups water
1 ham hock, or ham bone with some meat left on it (preferably a country-cured ham)
1 small onion
1/4 tsp. liquid smoke (unless the ham has been smoked, in that case, leave out liquid smoke)
1/4 pound butter
1 tsp. salt

In an iron skillet, melt the butter, add the chopped onion and simmer until lightly browned.

Put the beans and water in a large cooking pot, add the butter and browned onions and ham, along with remaining ingredients. Boil gently for two hours or until beans are tender.

Dried Apples

John Chapman, also known as "Johnny Appleseed," was a missionary who traveled across Ohio and Indiana in the early part of the 19th century, preaching the gospel and planting apple seeds. Apple seeds are unique in that they don't come "true" to the parent apple, and produce interesting variations. The McIntosh apple was found growing in Dundas County, Ontario, the Winesap was found in Rhode Island. Early apple varieties were crossed, bred and grafted into newer varieties. Apple seeds were planted in Vancouver, Washington as early as 1817, and many Indian tribes had adapted apples from their white neighbor's orchards, spreading this fruit widely by the time of the war between the states.

Drying apples made them more available and portable. Apples were peeled, sliced into 1/8 inch slices and layed out on tables to dry in the sun in trays. Cloth was often spread over the apples to keep off bugs (and to keep birds from eating the slices). The trays were taken inside at night to avoid dewfall, returned to the sunshine during the day. The apples would be dry in about a week, then stored in jars in the cellar or pantry. Dried apples were used for dried apple cakes, dried apple pies, cooked apples, as well as journey food.

Leather Britches

There are several recipes for using beans dried this way. The name comes from the way the whole bean pods look as they hang strung on strings to dry. The method and recipes came from several Indian tribes, including the Cherokee, who taught early settlers how to preserve this native American vegetable.

Leather britches are made from stemmed, whole green beans, in the snap-bean stage. With a large needle and string, run the needle through the top of each bean, making a string of beans. Hang strings of beans in a shady, airy place to dry (out of direct sun, such as in a barn loft or attic of the house). In about two months the beans should be dry. Many families left the beans hanging until ready to use, while still others put the beans, pods and all, into storage containers to protect from mice and moisture.

To cook: soak beans overnight and drain off water. Cover with boiling water in cooking pan, add scraps of meat (squirrel, pork, venison, whatever is available). Cook until tender - usually all day on a slow fire.

Other references call for drying the strung green beans in the sun for two months. To cook they were soaked for one hour in water, salt pork was added, bring to a boil. Reduce heat and simmer slowly, stirring occasionally, for three hours. More water is added as needed.

Hominy

Hominy could be made from any kind of shelled corn, although many families grew special hominy corn for the purpose.

For each quart of shelled corn, dissolve 2 tablespoons of concentrated lye in 1 gallon boiling water. (Lye was made from wood ashes; my grandfather kept a lye hopper - wood ashes were dumped into it from the wood stove, water was poured through and the water that came through was caught and saved. The lye sediment in the bottom was the object, and the water, as it cleared, was siphoned off the top, leaving concentrated lye in the bottom).

This concentrated lye, in the boiling water, would have the corn added, where it would be boiled until the hulls loosened - about 30 minutes. The corn was rinsed through several changes of fresh water to

8

remove the lye, then rubbed on a washboard or by hand, to remove the hulls. The hominy would then be left to soak in fresh water for a couple of hours, the water changed and the hominy washed again.

To cook the hominy, cover with boiling salt water (1 teaspoon salt to each quart hominy). Cook until almost tender and serve (or later the hominy would be canned in jars at this point, then given a hot-water bath to seal the jars). Canning by hot water bath isn't considered safe for hominy, now.

Fried Hominy

Hominy was often, but not always, made from hominy corn, hulled in a lye solution, then dried to preserve for later use. To cook this dried hominy, it would be soaked overnight, drained and simmered for several hours. Fried hominy was then made from this cooked hominy, or from left-over boiled hominy.

2 pounds hominy (5 cups cooked hominy)
salt pork or bacon (5 slices bacon will work)
4 or 5 wild
onions or 4 or 5
walking-onion
bulbs (or 1 small
onion, cut up)
water
salt and pepper

In a large iron skillet, fry the salt pork or bacon briefly with the cut up onions. Add hominy, salt and pepper and enough water to keep the hominy simmering. Simmer in this way in the fat, adding only small amounts of water, for about 30 minutes, or until tender. Serve with cooked greens or poke in season.

Poke Greens

In spring, pick poke shoots before leaves are fully opened - they look like large asparagus shoots. Snap off at ground level, collecting a basketful. Drop whole shoots in a pot of boiling water, bring to a boil again, drain and add fresh water. Bring to a boil a second time, and again pour off water, adding fresh. In this water add some salt pork or bacon drippings, along with salt. Boil for about twenty minutes. Good with cornbread (which is especially good to soak up the potlicker, or juice of the poke).

Fried Sweet Potatoes

Peel large sweet potatoes, slice in 1/4 inch slices. Fry in a skillet of meat fryings (bacon grease, lard). Cook slowly until the slices are beginning to brown. Turn over slices and cook until that side is browning. Add a sprinkle of salt and add brown sugar or molasses, continuing to simmer until tender (add a bit more grease or a small amout of water to keep from burning if needed).

Mint Julep

Confederate General Albery Sidney Johnston is often credited with first concocting this drink, although many sources claim that lots of Southern families had similar recipes and that General Sidney's is simply a variation. There are as many recipes for Mint Juleps as there are roads in Dixie, and much earlier references to the drink can

be found in recipe collections.

General Sidney's preference was that a julep be made by bruising a single leaf of catnip in a glass with a spoon. Fill the glass with cracked ice, pour whiskey into the glass to half an inch of the top.

Dissolve 2 tsp. sugar in a seperate glass of water and pour this into the whiskey-ice glass to fill up to the top. Add 2 or 3 good sprigs of fresh catnip to the glass and serve.

Another recipe calls for a frosted or frozen drink glass, filled with crushed ice and good bourbon. Garnish with a good sprig of mint "but never, never add sugar to bourbon."

And while the man in the field might have never been treated to this fine drink during the war, some officers certainly were. Gentlemen didn't always forget their manners in war, nor did those who were required to host them forget how to be good hosts. Mint juleps can be traced at least as far back as George Washington's mother!

References by Mary Ball Washington, in her own collection of recipes includes references to serving mint juleps with the gingerbread she preferred to serve with the drink. She named her gingerbread after General Lafayette, who served as a volunteer in the Continental army in the American Revolution (and died in 1834), and who especially loved Mrs. Washington's gingerbread. This demonstrates how recipes gained prominence, were changed, improved upon and passed along in later wars. Both the mint julep and the gingerbread recipe were common by the time of the War Between the States.

Communities would often turn out to help the wounded when battles were fought nearby, bringing food and supplies to help the tired soldiers, and carrying baskets of food to soldiers where they camped. This gingerbread recipe was in circulation at the time of the war and would likely have shown up in baskets of food prepared and brought to the area of the battleground.

Here is Mary Washington's recipe, much copied by North and South alike:

Lafayette's Gingerbread

1/2 c. butter
1/2 c. packed brown sugar
1 c. molasses
1/2 c. warm milk
2 T. ginger
1 tsp. cinnamon (or a mixture of cinnamon, mace and nutmeg)
1/2 c. brandy
3 c. sifted flour
3 eggs, beaten
1 orange
1 tsp. baking soda
1 T. warm water
raisins

Cream shortening, add sugar, beating until light. Stir in molasses, milk, spices, then brandy. Add eggs and flour, beating until smooth. Add grated orange rind, then stir in the juice. Add baking soda dissolved in water; beat until very light. Add raisins last. Bake in shallow pan (12 x 16) in 375 degree oven for 30 minutes. Serve warm with whipped cream and mint juleps.

Mrs. General Sheridan's
Patience Cakes

(This was more likely to have been served to her, than to have been her own recipe; the recipe, however, was named in her honor and shows up as late as 1891 in the famous Delmonico's Restaurant recipes)

Place in a vessel half a pound of powdered sugar, half a pound of flour, crack in three and a half eggs—no more and no less—and pour in a teaspoonful of anisette essence, and then with the spatula thoroughly mix for ten minutes. Slide a tube into the pastry bag, transfer the preparation into it; butter and flour two pastry baking pans; take hold of the bag and press down the preparation into each pan, into small round forms the size of twenty-five-cent pieces, keeping them half an inch apart from one another, and when done lay on shelves, in a dry place, for twenty-four hours. Place them in the hot oven to bake for fifteen minutes; take from out the oven, lay the pans on a table, and let cool off for thirty minutes; place them in a glass jar and serve when desired.

Root Beer
(Fermented only long enough to create the "fizz," these were non-alcoholic drinks)

Many families made their own root beer, and it was available in some towns in taverns. Made in the fall, generally, it was stored in the cellar in tightly-corked bottles. Some soldiers, when looting abandonded homes, came across this sweet treasure.

To make ten gallons of root beer, take
3 pounds burdock root, cut up
1 oz. essence of sassafras (or 1 pound sassafras root,

cut into 1 inch pieces)
1/2 pound hops
1 pint whole, dry, shelled corn, roasted brown

Boil all together almost an hour in 2 gallons of water. Strain while still hot into keg, adding enough cold water to make 10 gallons. Add molasses, honey or syrup to sweeten

(about 3 pounds sugar or 1 gallon honey). Add as much yeast as will raise a batch of 8 loaves of bread (about 1 cake yeast). Place keg in cellar, put bung loosely in keg and in 48 hours the root beer is ready to bottle. Make sure corks are wired down on the bottles after filling. Ready to drink in about a month, but better with age. Best to cool well before uncorking, to reduce the fizz from spewing the root beer out the top when uncorked.

A Second Root Beer Recipe

Water, 10 gallons; heat to 60° Fahrenheit, then add: 3 gallons of molasses; let it stand 2 hours, pour into a bowl, add powdered or bruised sassafras (roots) and wintergreen bark, of each half a pound;
yeast, 1 pint (probably liquid yeast)
bruised sarsaparilla root, half pound;
add water enough to make 25 gallons in all.
Ferment for 12 hours, then bottle.

Ginger Beer, *also non-alcoholic*

Ginger beer was made from wild ginger growing in shady moist places, found across the Ozarks, Missouri, Arkansas, Tennessee, Kentucky, Virginia and the Carolinas. The method is similar for Root Beer, above.

Take five and a half gallons water; add three-quarters of

14

a pound ginger root, bruised;
tartaric acid, half ounce;
white sugar, two and a half pounds;
whites of three eggs well beaten;
ten small teaspoonfuls of lemon essence;
yeast, one gill; (a gill equals 1/4 pint).

Boil the root for thirty minutes in one gallon of water; strain off, and put the essence in while hot; mix, make over night; in the morning, skim and bottle, keeping out the sediments.

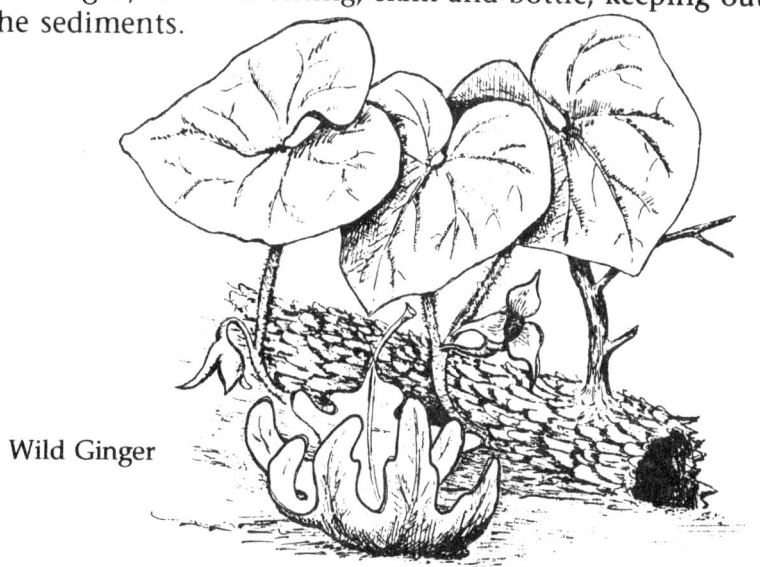

Wild Ginger

Spruce or Ginger Beer

Spruce beer was more popular in areas of the country such as Pennsylvania, Ohio, New York and adjoining states. Spruce beer is somewhat like root beer, a refreshing beverage to bring up out of the cellar.

Cold water, ten gallons;
boiling water, eleven gallons;
mix in a barrel; add molasses, thirty pounds, of brown sugar twenty-four pounds;
oil of spruce, or any oil of which you wish the flavor, one ounce; add one pint of yeast; ferment. Bottle in two or three days.
If you wish white spruce beer, use

seventeen ounces ginger root bruised, and a few hops; boil for thirty minutes in three gallons of water, strain and mix well; let it stand two hours and bottle, using yeast, of course, as before.

This recipe calls for mixing the boiling water and cold water together. Without doubt this was a mistake in the original recording of the recipe, as there would be no purpose to that method. Adding the sweetening agent to boiling water, however, would insure its being dissolved. The recipe could be made in much smaller batches, of course.

Hopbeer

These carbonated non-alcoholic beverage recipes differed from one part of the country to the other, depending upon what materials were available and what each family liked. My own ancestors made only Root Beer, while other families record Hopbeer or Gingerbeer. For the thirsty soldier, marching a long way from home, any drink offered him on the road was welcomed.

Mix 14 pounds of molasses, and 11 gallons of water, well together, and boil them for 2 hours with 6 ounces hops.

When quite cool, add a cupful of yeast, and stir it in well by a gallon or two at a time.

Let it ferment for 16 hours in a tub covered with a sack; then put it in a 9 gallon cask, and keep it filled up; bung it down in 2 days, and in 7 days it will be fit to drink, and will be stronger than London porter.

(Keeping the cask "filled up" refers to the action of the fermentation - if made in warm weather the escaping gases might cause some of the liquid to bubble out the bung hole, in which case a little more water would be added).

Another Carbonated Hopbeer Recipe:

Hops, 6 ounces
Molasses, 5 quarts; boil the hops until the strength is out, strain into a 30 gallon barrel. Add the molasses and 1 teacupful of yeast, then fill up with water; shake it well and leave the bung out until fermented, which will be in about 24 hours. Bung it up and it will be fit to use in about 3 days.

Molasses Beer

Hops, 1 ounce
Water, 1 gallon
Boil for 10 minutes; strain and add:
Molasses, 1 pound. When lukewarm, add yeast, 1 spoonful. Ferment as in recipe above.

Lemon Beer

Carbonated Lemon "Beer" was popular along coastal areas and at river towns where lemons were accessible and several families recorded this early recipe.

To make 20 gallons, boil 6 ounces of bruised ginger root, a quarter pound of cream of tartar, for about 30 minutes in 2 gallons of water. This will be strained in 13 pounds coffee sugar (white sugar), on which you have put half an ounce of oil of lemon and 6 good (fresh) lemons squeezed up together, having warm water enough to make the whole 20 gallons just so hot that you can hold your hand in it without burning, or about seventy degrees of heat; put in one and a half pints of hops or brewers yeast, worked into paste with 5 or 6 ounces flour. Let it work over night, then strain and bottle for use.

Note: use caution with this one, as is true of the above recipes - a lot of "fizz" will build up in the bottled brew. The yeast or hops gives the liquid in all the recipes a natural carbonation (rather than making an alcoholic drink). This carbonation will pop the corks in a loud thunderstorm if you don't have them wired on tight. Additionally, be sure to cool the bottle well - several hours - before uncorking or it will spew the contents out like a bottle of champagne that's been shaken.

Sassafras Tea

Dig sassafras roots during winter or early spring, before "sap's a'risin." Wash well, cut root into short pieces - 1 or 2 inches long. Drop 2 or 3 of these pieces into a quart of boiling water and simmer for 5 minutes. Pour into cups, adding honey and milk if available.

The roots were dried and used the year around, and sassafras tea was sometimes used as a flavoring, like we use vanilla or extracts today. The dried roots were easy to carry in the pocket and the tea was a welcome smell and taste on cold winter nights, or as a wake-up beverage on frosty mornings.

Persimmon Tea

In summer, gather fresh green leaves of the persimmon tree. Use 4 or 5 leaves for a pot of tea, pouring boiling water over the leaves in the warmed pot. Steep for a few minutes and pour into mugs. Good as a cold tea. The leaves could be dried and used in winter, as well.

Sumac Tea

This was an Indian beverage and settlers learned that it was especially good as a thirst-quenching summer beverage. Cut one or two red seed heads of sumac (poison sumac has white berries and is a contact poison, like poison ivy - red berried sumacs are tasty and tart). Cut away as much of the stem as is easily done and drop the seed clusters into a kettle of boiling water (about 5 cups). Remove from heat, cover pot with lid and let steep a few minutes.

Strain, add honey and serve. The tea is like a hot lemonade, and is also good cold. Spices, like cardamom, a clove or spice bush berry were often added with the seed clusters. Soldiers who knew sumac would pick the sticky berries to suck on to quench their thirst when traveling long distances.

Corn Pone

This could be cooked around a small campfire when away from the main unit. Everyone who cold get cornmeal, carried some for such emergencies.

Mix about 2 cups corn meal with some salt. Add meat fryings and about a cup and a half water (boiling if possible). Add a pinch of clean wood ashes for leavening.

Stir together, and with the hands, make out into small balls, or pones. Flatten this stiff batter just a little and lay out on a clean, hot rock near the fire. Turn as necessary and cook until done - about twenty to thirty minutes, depending upon the heat of the fire.

Vinegar pies were common, being the fore-runners of our lemon pies. There are lots of variations, but this one was named **Gatlinburg Vinegar Pie**, *and I pass it along as is:*

3 eggs, beaten
1 1/2 cups melted butter or mild lard
3 tablespoons dark cider vinegar
1 cup sugar
Add melted butter to beaten eggs, then the vinegar and sugar. Pour into unbaked crust and bake in a very hot oven (450°) for ten minutes, reduce heat to 300° and bake until pie is set or knife comes out clean when inserted (about 35-40 minutes).

Cracklin' Bread

Cracklin's are what's left of the hog fat after the lard has been cooked or rendered out. It's somewhat like fried pig skins, but with tiny bits of tenderloin still hidden inside. Cracklin's were used as a snack, and mixed in a variety of foods, into mush and as seasoning.

2 cups corn meal	1 egg
1 tsp. salt	1/2 cup cracklin's
3 tsp. baking powder	1/2 cup milk
1/2 cup water	

Mix ingredients, except cracklin's, until smooth. Stir in the cracklin's, bake in a well-greased iron skillet for 45 mintues in 350° oven.

Plantation Biscuits

1 cup flour   1/2 tsp. salt
1 1/2 tsp. baking powder   2 T. shortening
1/2 cup buttermilk

Sift flour, salt and baking powder. Cut in shortening until the consistency of corn meal; add buttermilk, mix briefly and roll out on floured board. Cut with tin can or biscuit cutter, laying close together on baking pan. Bake in 375° oven for about 20 minutes, or until just browning on top.

Possum and Yams

Dress the possum, removing as much of the fat as possible (a young possum is best). Wash thoroughly inside and out. Parboil until tender - a couple of hours. Drain and place in a baking dish, along with an onion, salt and pepper, adding 4 whole sweet potatoes, or yams. Bake until potatoes are done, basting every 15 minutes with drippings.

Fried Squirrel

Squirrel was easily found by small groups of soldiers, or a single soldier on an errand. A squirrel might be killed by "barking," a method of shooting that caused the least amount of damage to the meat. Barking a squirrel was shooting at the tree, just beneath the squirrel so that the exploding bark would hit the squirrel with such force, under its chest or belly, that it killed the animal, or at least knocked it unconscious, where it fell to the ground to be quickly picked up and dressed. Good marksmen commonly barked squirrels this way, bringing home several dead squirrels without a mark or wound on them to show how they had been killed.

After the squirrel is dressed, some people preferred to singe the meat briefly to remove any hair left by the skinning. Wash the squirrel, inside and out in fresh water. Cut apart, at the joints, and halve the ribs from the back portion. This gives you 4 good pieces and 2 bony pieces. Dredge in flour, salt and pepper having been added.

Fry in a medium-hot skillet in which you've added about a cup of lard or bacon drippings; fry until browned on one side; turn and repeat on the other side, being careful that the fire isn't so hot that the meat burns or cooks too fast (which will cause the meat to be tough). If this is a young squirrel, the meat will be ready to eat.

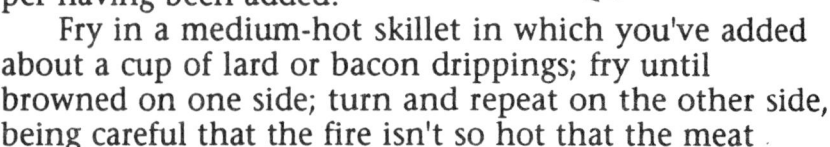

'On the other hand, if the squirrel was an old one, it will be tough and require more cooking. Leave the squirrel in the cooking pan, and add 2 cups of water and cover with a lid, and let it simmer slowly for about an hour or more (turn occasionally and add more water). Near the end of cooking time, add about a cup of additional water, stirring up the fryings in the bottom, add 4 tablespoons of flour and stir in well. Let boil briefly until this gravy is thickened and serve with biscuits.

For portions, one man can easily eat a squirrel, if there's little else to eat. If other food is served, such as plenty of biscuits, gravy, some potatoes or greens, one squirrel should feed two or three men, but it means only one squirrel leg each and a fight over the ribs.

Rebel Stew

This recipe is adapted from many sources, and although it isn't directly authentic to the time, the kind of stew, the method of making it, along with the ingredients, are all very true to the time of the War Between the States. Many variations exist, many soldiers have made it on countless battlefields throughout history, and remains today in our own cooking practises as "Camper's Stew" or "Trail Stew."

1 pound, more or less, of beef, pork, venison, squirrel or whatever meat is available
1 small onion, chopped
4 potatoes, cut up
3 carrots, cut up
1 cup corn, hominy, dried or parched corn
1 cup okra, if available
3 cut up fresh tomatoes (or 1 pint canned)
Seasoning (salt & pepper, certainly, a chili pepper or sweet pepper, or dash of tobasco sauce, whatever might be handy. *Southern and Ozarks soldiers who knew spicebush that grew along streams, knew to use the leaves, twigs or berries of that plant for the best seasoning for the stew)*

In an iron kettle, brown the meat and onions in bacon drippings; add the rest of the ingredients, pour in about a gallon of water and let simmer for a few hours, or overnight, with a lid on the pot. Awhile before serving, add more water if needed, build up the fire and let stew come to a boil. Mix 1/2 cup flour or finely-ground cornmeal in 2 cups water and stir to mix well; add to boiling stew and stir until slightly thickened, adjust salt and pepper and serve.

My Great-grandfather, George Washington Garrison, was born before the Civil War. As a young man he planted a large orchard in western Missouri. Great-Grandpa grafted, crossed and improved apple varieties, growing many kinds and came to be known in surrounding counties for his outstanding apple harvests. People came to his farm in teams and wagons to buy his fine apples every fall. The following recipe comes to me through my mother, who learned the recipe from my Grandma Harper, who was one of twelve of Great-Grandpa's and Great-Grandma's children. These apple dumplings have been served since at least the time of the Civil war, and probably well before.

Directions for Apple Dumplings

Peel and slice 4 apples. Make pie crust pastry and divide into 4 pieces. With one piece of pastry, roll out thin and pile one-fourth of the sliced apples in the center. Add 1 Tbsp. sugar, 1/2 tsp. cinnamon and 1/2 tsp. nutmeg. Bring pastry over and press edges together. Place upside down (seam-side down) in deep pan or casserole. Repeat with remaining pastry and apple slices.

Dissolve 1 cup sugar and 4 Tbsp. butter in 4 cups boiling water. Pour over dumplings and bake in medium oven (350°) for about an hour. If sauce gets too low, pour in a little more water during baking. These are delicious served hot with thick cream poured over, or served cold with hot tea.

Treatments for Horses

Keeping horses in good shape was nearly as important as keeping the soldier going. Each man knew his horse's needs and tended him daily. When injuries occurred, the scratch, cut or swelling was attended to rapidly. Here are some methods used.

White Liniment

Used to treat wounds on horses and humans, too.

1/4 pint salty meat grease
1/2 pint turpentine
1/2 pint kerosene
1 pint vinegar
1 pint hard apple cider
handful of salt
3 or 4 egg whites

Mix well and store in jar. This was used for cuts and bruises. Horses would often suffered cuts and scratches while being ridden through brush or in battle. Every man knew the importance of tending the animal's wounds quickly.

For cuts on horses, especially from wire or metal: Ground root of goldenseal mixed with an equal amount of lard and applied. The goldenseal is a drawing agent and the lard kept the wound soft. Pine tar was also used on cuts of this type with good effect.

A remedy for swelling on horse or man, was to apply a hot poultice of dried comfrey mixed with lard. This was left on for an hour or more, the poultice reheated and applied again. Swelling soon disappeared.

To deter biting flies on horses - Soldiers would sometimes tie a leafy branch to the horse's throat latch. The movement of the horse made the branch sway, keeping flies from biting the front and legs of the horse. A rag soaked in kerosene was sometimes attached to the throat latch, under the bridle, for the same purpose.

Treating a horse for distemper often consisted of burning old leather under the horse's nose. The smoke would keep the horse's nose from clogging, thus preventing choking.

A common method for relieving a horse of gas was to make him jump logs. It was necessary to keep the horse from lying down, which would generally result in the horse's death.

To prevent snow water from penetrating shoes: This simple and effectual remedy is nothing more than a little beeswax and mutton suet, warmed in a pipkin until it is liquid. It was then rubbed over the edges of the sole and where the stitches are, repelling the dampness.

One remedy in my own family for **treating neck sores on horses** from the friction of the collar as is rubbed on the-horse's necks as they worked, was to apply green leaves of jimpson weed. My grandfather claimed it worked.

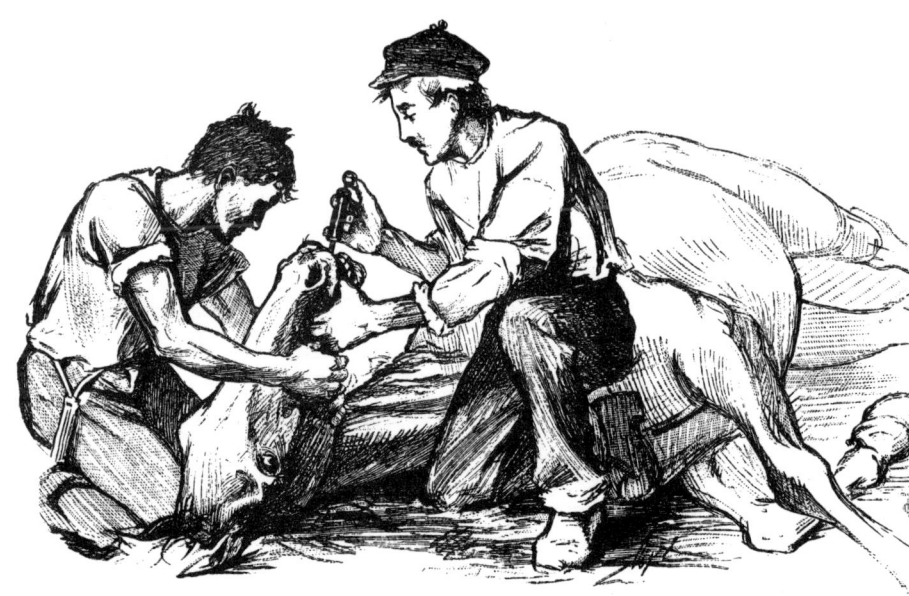

Bibliography

Filippini, Alessandro, *The Table: How to Buy Food, How to Cook it and How to Serve it*, Delmonico's Restaurant, New York, 1889.
Garrison family recipes, unpublished collection, circ. late 1800s.
Herter, Leonard, *Bull Cook and Authentic Historical Recipes*, 1963.
Long, Mada H., private collection, 1913-1987.
Massey, Ellen Gray, *Bittersweet Country*, 1978.
Phelan, F.E., *Young's Great Book of Secrets*, circ. 1880.
Sharpe and Underwood, *American Indian Cooking & Lore*, Cherokee, 1973.
Slagle, Elias, The Diary of E. Slagle, unpublished collection, 1859.

Miscellaneous Foods in Use During the War:

Hardtack: A hard biscuit made only with flour and water. Hardtack was made in enormous quantities by volunteers and later by manufacturers and shipped to armies in the field. The hard "bread" could be carried for months without spoiling, although it often molded or harbored bugs. Sometimes a bit of wood ash was mixed with the flour and water as it was made, giving a small amount of leavening, but generally not. The hardtack was crumbled in soup, boiled in water, pounded with rocks and mixed with any edible liquid in order for the soldier to survive another day. The idea of portable food was good but in actuality hardtack was horrible and akin to eating plaster.

Miscellaneous Foods in use during the War

Canned Food - Tinned foods, preserved in tin cans, were fully in use by the mid-1800s. These canned foods weren't always safe, due to processes that sometimes didn't work well. Additionally, the tins were not coated with non-corrosive metal as they are today, so flavors often deteriorated. The contents would sometimes dehydrate, acidic foods would corrode the metal and the food could become rancid, or worse, poisonous. Still, tinned foods made travel over long distances easier and miners, trappers and other long-distance travelers had been using these processed foods since about 1850. The common soldier made use of tinned foods when they were available, sometimes buying them in towns, but more often than not, those more expensive items were not to be found.

Tomatoes - There is the false belief that tomatoes were not in general use at the time of the Civil War. In actuality, the tomato as a vegetable, was common. Thomas Jefferson grew them on his plantation at Monticello at the end of the 1700s. Records from the Civil War show townspeople riding out in wagons to the site of on-going battles during the War, carrying with them pies, bread, beverages and garden crops, including tomatoes, to hand out to the tired and hungry fighting men.

Ice Cream - Ice cream isn't an accomplishment of the 20th century. One recipe from the mid-1800s describes putting a layer of broken ice and salt in a tub. The cream mixture was then put in a tall container with a lid, set in the tub and surrounded with alternating layers of ice and salt. A woolen blanket was pinned over the top of the cream container and it was moved by hand inside the ice-salt mixture. Every few minutes the blanket was removed and the frozen cream scraped away from the sides of the container and mixed into the freezing liquid. With the lid returned and the blanket spread back over, the freezing would continue, scraping down the sides again and again until the entire mixture was frozen. It was a lot of trouble, but ice cream has always been a real treat and one worth waiting for. The hand-cranked, mechanical ice cream freezer was invented by a woman some years after the Civil War, making this arduous job easier.

ᏆIndex to RecipesᏆ

ᏆIndex to People ListedᏆ

Other books by Jim Long

• Herbal Medicines of the Santa Fe Trail   4.95
• Just for Men, Body Preparations from the Garden   4.95
• Making Herbal Cosmetics   4.95
• Planning & Hosting Successful, Profitable Festivals   4.95
• Herbs, Just for Fun, a Guide to Growing & Using Herbs   3.95
• Successful Self-Publishing   4.95
• Tea & Cakes Under the Trellis - a Beginner's Guide to
  Making Bentwood Trellises, with Recipes   4.95
• How to Make Romantic Bentwood Trellises,
  Gates, Fences & Arbors   4.95
• Free Publicity for You, Your Organization or Your Business   4.95
• Classic Seasoning Blends from Your Garden   4.95
• Dream Pillows & Potions   3.95
• Making Profits from Dream Pillows   9.95

And these books by Jim Long,
available from Storey Publishing,
or directly from the author at the address in the front of this book:

• Making Bentwood Trellises, Arbors, Gates & Fences   19.95 + $3 postage
• Making Herbal Dream Pillows   14.95 + $3 postage

Visit the author's website:
www.longcreekherbs.com